梁燕 編著

「新手入廚系列」

百變雞翼

前言

以往，農村以耕種度日，生活比較簡樸。每年，五穀豐收，春節將至，家家都會劏雞敬神。在人們的心目中，「雞」的份量相當重要。

現在大部分人都吃過雞翼，是雞不可或缺的美食部分，雞翼不僅繼承了雞的優點，而且更將其發揚光大。雞翼肉質軟滑，製作簡易，烹調方法蒸、煮、炸、釀⋯⋯多彩多姿，滋味無窮，令人垂涎三尺。

雞翼主要由三部分組成：「雞槌」、「雞中翼」和「雞翼尖」，聰明的商人將其分拆獨立售賣。在烹製過程中，「雞中翼」更容易掌握火候，入口更為方便，而「雞槌」及「雞翼尖」也各有支持者。

每逢假期，在郊野公園的燒烤場內，雞翼成為必然的主角，越來越多的人喜歡上它，真要多謝這一群燒烤的擁護者。現在，幾乎所有的食店都會提供雞翼的菜式。至於居家烹製，大部分家庭主婦也都會在雪櫃中準備一份生雞翼以備不時之需：搭配不同的醬汁或食材製作成數十種不同味道和款式的雞翼，真可謂濃情美意，百味綻放！

目錄

雞翼的處理方法

法雞翼快速解凍法

1. 用自來水直接沖洗雞翼，可令雞翼快速解凍。
2. 將雞翼浸於鹽水中，有助加速解凍。
3. 將雞翼放進微波爐內，按下解凍掣，即可迅速解凍。

雞全翼去骨法

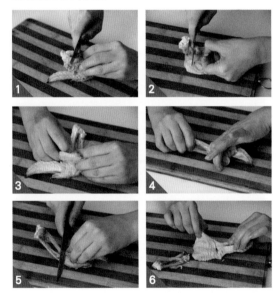

1. 雞全翼先解凍。切去雞槌部分，在雞翼頂部，即比較大的位置用小刀剝去雞翼的根。
2. 沿着雞翼的骨向下剝。
3. 將雞肉連皮向下推。
4. 將雞翼的肉和骨用刀切開。
5. 保留雞翼尖部分，將雞翼的肉從外向內翻入。
6. 洗淨雞翼後瀝乾水分，即可加入醃料。

雞中翼去骨法

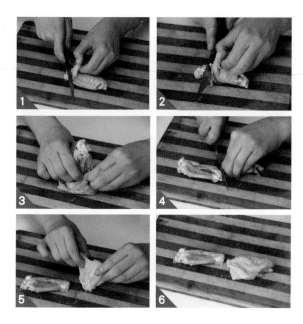

1. 雞中翼先解凍。在雞翼頂部，即比較大的位置用小刀剝去雞翼的根。
2. 沿着雞翼的骨向下剝。
3. 將雞肉連皮向下推。
4. 將雞翼的肉和骨用刀切開。
5. 將雞翼的肉從外向內翻入。
6. 洗淨雞翼後瀝乾水分，即可加入醃料。

去除雪藏味的方法

醃雞翼時加入適量酒、胡椒粉拌勻，便可去除雪藏味。

非洲辣雞翼

African Spicy Chicken Wings

◎◎ 材料 | Ingredients

雞中翼 8 隻
紅蘿蔔 1/2 個
薯仔 1 個
洋蔥 1 個
乾蔥 1 粒
紅辣椒 2 隻

8 chicken mid–joint wings
1/2 carrot
1 potato
1 onion
1 clove shallot
2 red chilies

⦾ 醃料 | Marinade

生抽 1 茶匙
糖 1/2 茶匙

1 tsp light soy sauce
1/2 tsp sugar

⦾ 汁料 | Sauce

黑胡椒 2 茶匙
鹽 1 1/2 茶匙
水 1/2 杯

2 tsps black pepper
1 1/2 tsps salt
1/2 cup water

⦾ 做法 | Method

1. 雞翼洗淨，瀝乾水分。
2. 紅蘿蔔、薯仔分別洗淨，去皮切粒；洋葱去衣，洗淨，切粒。
3. 燒熱油鑊，下雞翼煎至金黃，盛起隔油備用。
4. 燒熱油鑊，爆香乾葱、洋葱和紅辣椒，下汁料煮 15 分鐘，最後放入薯仔、雞翼，煮至薯仔略腍即可。

1. Rinse chicken wings well and drain.
2. Wash carrot and potato well, peel and dice. Wash onion well, remove skin and dice.
3. Heat wok with oil, fry chicken wings until golden. Dish up and drain excess oil, set aside.
4. Heat wok with oil, stir-fry shallot, onion and red chilies. Add in sauce and cook for 15 minutes. Put in potato and chicken and cook until potato is done.

三杯雞槌

Three Cups Drumsticks

ⵔⵔⵔ 材料 | Ingredients

雞槌 8 隻	8 drumsticks
蒜片 5 粒量	5 cloves sliced garlic
薑 2 片	2 slices ginger
紅辣椒絲 1 隻量	1 shredded red chili
九層塔 10 片	10 pieces basils

2~4人
Serves 2~4

20~25 分鐘
20~25 minutes

⊗⊗ 調味料 | Marinade

酒 1/2 杯	1/2 cup wine
生抽 3 湯匙	3 tbsps light soy sauce
麻油 3 湯匙	3 tbsps sesame oil
糖 1 茶匙	1 tsp sugar
鹽 1/2 茶匙	1/2 tsp salt
水 1/2 杯	1/2 cup water
胡椒粉少許	pepper powder

⊗⊗ 做法 | Method

1. 雞槌洗淨，瀝乾水分，汆水。
2. 燒熱油鑊，下雞槌炸至金黃，盛起隔油備用。
3. 燒熱油鑊，爆香薑片、蒜片和紅辣椒絲，加入雞槌和調味料炒勻至汁料濃稠，加入九層塔炒勻即可。

1. Rinse drumsticks well, drain and scald.
2. Heat wok with oil, deep-fry drumsticks until golden, dish up and drain excess oil, set aside.
3. Heat wok with oil, stir-fry ginger, garlic and red chili until fragrant. Add in drumsticks and seasonings, stir-fry until sauce thickens, add in basil and stir well, serve.

紐約香辣黑椒雞翼

Baked Black Pepper Chicken Wings in New York Style

⟨⟨⟨ 材料 | Ingredients

雞中翼 8 隻
辣椒仔辣汁 1/2 杯
牛油 4 湯匙

8 chicken mid–joint wings
1/2 cup Tabasco
4 tbsps butter

2~4 人
Serves 2~4

20~25 分鐘
20~25 minutes

⓪ 調味料 | Seasonings

麵粉 1 杯
蒜粉 1 茶匙
黑胡椒粉 1 茶匙
鹽 1/2 茶匙

1 cup flour
1 tsp garlic powder
1 tsp black pepper powder
1/2 tsp salt

入廚貼士 | Cooking Tips

- 辣椒仔辣汁可以 Hot Sauce 代替。
- Tabasco could be replaced by Hot Sauce.

⓪ 做法 | Method

1. 雞翼洗淨，瀝乾水分。
2. 預熱焗爐至 200℃。
3. 將調味料混合，把雞翼均勻地沾上調味料，排放在已掃油的焗盤上，焗 20 分鐘至金黃色。
4. 以慢火融化牛油，加入辣醬，再加入焗好的雞翼拌勻，讓辣醬均勻沾在表面即成。

1. Rinse chicken wings well and drain.
2. Preheat an oven to 200 ℃.
3. Mix all the seasonings well and dip chicken wings with seasonings evenly. Arrange onto a greased baking tray and bake for 20 minutes until golden.
4. Melt butter over low heat, add in Tabasco. And stir well with cooked chicken wings until the sauce is evenly coated, serve.

檸汁青芒川辣雞翼

Spicy Chicken Wings with Lemon Juice and Thai Mango

雞中翼 8 隻
泰國青芒果 1 個
車厘茄 10 粒

8 chicken mid-joint wings
1 Thai mango
10 cherry tomatoes

2~4 人
Serves 2~4

20~25 分鐘
20~25 minutes

火辣辣
Spicy

◯◯◯ 醃料 | Marinade

花椒粉 1 湯匙	1 tbsp Chinese red pepper powder
黑胡椒醬 1/2 湯匙	1/2 tbsp black pepper paste
麻油 1/2 湯匙	1/2 tbsp sesame oil
糖 2 茶匙	2 tsps sugar
鹽 2 茶匙	2 tsps salt

◯◯◯ 調味料 | Seasonings

紅辣椒（切碎）1 隻	1 chopped red chili
魚露 1 1/2 湯匙	1 1/2 tbsps fish sauce
檸檬汁 1 1/2 湯匙	1 1/2 tbsps lemon juice
糖 1 湯匙	1 tbsp sugar

◯◯◯ 做法 | Method

1. 雞翼洗淨，瀝乾水分，加入醃料醃 30 分鐘。
2. 泰國青芒果洗淨，去皮，去核，刨絲。
3. 車厘茄洗淨，開邊。
4. 燒熱油鑊，將雞翼煎至金黃熟透，盛起隔油備用。
5. 燒熱油鑊，用慢火煮滾調味料，加入青芒果絲、車厘茄，淋在雞翼上，即成。

1. Rinse chicken wings well and drain. Marinate for 30 minutes.
2. Wash Thai mango well, removes skin and seed, shred.
3. Wash cherry tomatoes well, cut into halves.
4. Heat wok with oil, fry chicken wings until golden and done, dish up and drain excess oil, set aside.
5. Heat wok with oil, bring seasonings to a boil over low heat, add in shredded Thai mango and cherry tomatoes. Pour onto chicken wings and serve.

2~4 人
Serves 2~4

15~20 分鐘
15~20 minutes

花椒玫瑰露雞翼

Chicken Wings with Red Pepper and Rose Wine Sauce

⊂⊃⊃ 材料 | Ingredients

雞中翼 8 隻
辣椒仔辣汁 3 湯匙
玫瑰露 1/2 杯
花椒 8 粒
薑 5 片
鹽 2 茶匙

8 chicken mid-joint wings
3 tbsps Tabasco
1/2 cup rose wine
8 cloves Chinese red pepper
5 slices ginger
2 tsps salt

⟨⟨⟨ 做法 | Method

1. 雞翼洗淨，瀝乾水分。

2. 燒滾水，加入薑和鹽，把雞翼浸至熟透，瀝
   乾水分，置大碗中備用。

3. 燒熱油鑊，爆香花椒，加入酒和辣椒仔辣汁
   拌勻煮滾，拌入雞翼中，待涼後，加蓋，置
   於雪櫃內醃 8 小時，即可。

1. Rinse chicken wings and drain.
2. Bring water to a boil, add in ginger and
   salt, soak chicken wings until done. Drain
   and put in a big bowl, set aside.
3. Heat wok with oil, stir-fry Chinese red
   pepper until fragrant, add in wine and
   Tabasco, stir well and bring to a boil,
   pour into chicken wings. Leave to cool,
   cover and put into the refrigerator,
   marinate for 8 hours and serve.

露斯瑪莉咖喱雞翼

Curry Chicken Wings with Rosemary

⬭⬭ 材料 | Ingredients

雞中翼 8 隻
紅蘿蔔 1 個
薯仔 1 個

8 chicken mid-joint wings
1 carrot
1 potato

⬭⬭ 醃料 | Marinade

酒 2 茶匙
糖 1/2 茶匙
鹽 1/2 茶匙

2 tsps wine
1/2 tsp sugar
1/2 tsp salt

⦿ 汁料 | Sauce

茄汁 8 湯匙	8 tbsps ketchup
紹酒 3 茶匙	3 tsps Shaoxing wine
老抽 3 茶匙	3 tsps dark soy sauce
生抽 2 茶匙	2 tsps light soy sauce
糖 2 茶匙	2 tsps sugar
白醋 2 茶匙	2 tsps white vinegar
咖喱粉 2 茶匙	2 tsps curry powder
鹽 1/2 茶匙	1/2 tsp salt
露斯瑪莉 少許	rosemary
胡椒粉 少許	pepper powder
水 1 杯	1 cup water

⦿ 做法 | Method

1. 雞翼洗淨，瀝乾水分，加入醃料醃 30 分鐘。
2. 紅蘿蔔、薯仔洗淨，去皮，切件。
3. 燒熱油鑊，將雞翼煎至金黃，加入紅蘿蔔、薯仔和汁料同燜至雞翼熟透即可。

1. Rinse chicken wings well and drain. Marinate for 30 minutes.
2. Wash carrot and potato well, peel and cut into pieces.
3. Heat wok with oil, fry chicken wings until golden. Add in carrot, potato and sauce, stew until chicken wings are done, serve.

酸辣貴妃雞翼

Stewed Chicken Wings with Hot and Sour Sauce

⊙⊙⊙ 材料 | Ingredients

雞中翼 8 隻
紅蘿蔔 1 條
洋蔥 1 個
蔥段 1 棵量
薑 2 片
蒜茸 1 茶匙
乾蔥茸 1 茶匙

8 chicken mid-joint wings
1 carrot
1 onion
1 stalk sectioned spring onion
2 slices ginger
1 tsp minced garlic
1 tsp minced shallot

2~4 人
Serves 2~4

20~25 分鐘
20~25 minutes

◯◯◯ 醃料 | Marinade

生抽 1 湯匙	1 tbsp light soy sauce
酒 1 茶匙	1 tsp wine
胡椒粉少許	pepper powder

◯◯◯ 汁料 | Sauce

茄汁 3 湯匙	3 tbsps ketchup
生抽 1/2 湯匙	1/2 tbsp light soy sauce
豆瓣醬 2 茶匙	2 tsps broad bean paste
糖 1 1/2 茶匙	1 1/2 tsps sugar
鹽 1/4 茶匙	1/4 tsp salt
水 1/2 杯	1/2 cup water
麻油少許	sesame oil

◯◯◯ 做法 | Method

1. 雞翼洗淨，瀝乾水分，加入醃料醃 30 分鐘。

2. 紅蘿蔔洗淨，切塊；洋葱洗淨，去衣，切塊。

3. 雞翼、洋葱分別泡油備用。

4. 燒熱油鑊，爆香薑、蒜茸、乾葱茸，雞翼回鑊，灒酒，加入汁料（麻油除外）、紅蘿蔔、洋葱和葱段，加蓋煮至汁液濃稠，下麻油拌炒勻即可。

1. Rinse chicken wings well and drain. Marinate for 30 minutes.

2. Wash carrot well and cut into pieces. Wash onion well, peel and cut into pieces.

3. Scald chicken wings and onion in warm oil respectively, set aside.

4. Heat wok with oil, stir-fry ginger, garlic and shallot until fragrant. Return chicken wings into wok, sizzle with wine, and add in sauce (except sesame oil), carrot, onion and spring onion, cover and cook until sauce thickens. Add in sesame oil and stir-fry, serve.

沙茶香雞翼

Baked Chicken Wings with Sa Cha Sauce

材料 | Ingredients

雞中翼 8 隻

8 chicken mid-joint wings

2~4 人
Serves 2~4

30~35 分鐘
30~35 minutes

◯◯ 醃料 | Marinade

蒜茸 2 湯匙	2 tbsps minced garlic
沙茶醬 2 1/2 茶匙	2 1/2 tsps Sa Cha sauce
鹽 1/2 茶匙	1/2 tsp salt
糖 1/2 茶匙	1/2 tsp sugar
酒 1/2 茶匙	1/2 tsp wine
胡椒粉少許	pepper powder

◯◯ 做法 | Method

1. 雞翼洗淨，瀝乾水分，加入醃料醃 30 分鐘。
2. 預熱焗爐至 200℃。
3. 將雞翼排放在焗盤上，置於焗爐內焗 30 分鐘即成。

1. Rinse chicken wings well and drain. Marinate for 30 minutes.
2. Preheat oven to 200℃ .
3. Arrange chicken wings onto a bake tray, put into oven and bake for 30 minutes, serve.

Spicy Chicken Wings with Twin
Sauce in Mexican Style

雙料墨西哥香辣雞

材料 | Ingredients

雞槌 5 隻
雞中翼 5 隻

5 drumsticks
5 chicken mid-joint wings

醃料 | Marinade

鹽 1/2 茶匙
酒 1/2 茶匙
胡椒粉少許

1/2 tsp salt
1/2 wine
pepper powder

◯◯ 汁料 | Sauce

(1) 墨西哥路易斯安那辣醬 1/2 杯
    溶牛油 2 湯匙
    蒜粉 1 茶匙

(2) 美式燒烤汁 5 湯匙
    辣椒粉 1/2 茶匙

(1) 1/2 cup Louisiana hot sauce
    2 tbsps melted butter
    1 tsp garlic powder

(2) 5 tbsps American barbecue sauce
    1/2 tsp chili powder

◯◯ 做法 | Method

1. 雞翼和雞槌洗淨，瀝乾水分，加入醃料醃 30 分鐘。

2. 燒熱油鑊，將雞翼和雞槌炸至金黃熟透，盛起隔油備用。

3. 拌勻汁料 (1)，將雞翼沾滿汁液，上碟。

4. 將雞槌沾上拌勻的汁料 (2)，上碟，即成。

1. Rinse chicken wings and drumsticks well and drain. Marinate for 30 minutes.

2. Heat wok with oil, deep-fry chicken wings and drumsticks until golden and done. Dish up and drain excess oil, set aside.

3. Stir well sauce (1), dip chicken wings with sauce (1), arrange onto the plate.

4. Stir well sauce (2), dip drumsticks with sauce (2), arrange onto the plate and serve.

入廚貼士 | Cooking Tips

- 墨西哥路易斯安那辣醬和美式燒烤汁可在大型超級市場購買。
- Louisiana hot sauce and American barbecue sauce could be bought from superstores.

蒜茸茄汁焗雞翼

Baked Chicken Wings with Minced Garlic and Ketchup

◯◯ 材料 | Ingredients

雞中翼 8 隻

8 chicken mid-joint wings

2~4 人
Serves 2~4

30~35 分鐘
30~35 minutes

醃料 | Marinade

茄汁 5 湯匙	5 tbsps ketchup
蒜茸 2 湯匙	2 tbsps minced garlic
糖 1 1/2 茶匙	1 1/2 tsp sugar
辣椒粉 1 茶匙	1 tsp red chili powder
酒 1 茶匙	1 tsp wine
鹽 1/2 茶匙	1/2 tsp salt
胡椒粉少許	pepper powder

做法 | Method

1. 雞翼洗淨，瀝乾水分，加入醃料醃 30 分鐘。
2. 預熱焗爐至 200℃。
3. 將雞翼排放在已掃油的焗盤上，焗 20 分鐘，再翻轉另一面焗 10 分鐘至金黃色即成。

1. Rinse chicken wings well and drain. Marinate for 30 minutes.
2. Preheat an oven to 200℃。
3. Arrange chicken wings onto a greased baking tray. Bake for 20 minutes, turn over and bake for another 10 minutes until golden.

風沙雞翼

Deep-fried Chicken Wings with Golden Garlic

⬤⬤⬤ 材料 | Ingredients

雞中翼 8 隻
蒜茸 300 克

8 chicken mid-joint wings
300 minced garlic

⟨⟨⟨ 醃料 | Marinade

鹽 2 茶匙
酒 1 茶匙

2 tsps salt
1 tsp wine

⟨⟨⟨ 調味料 | Seasonings

鹽 3 茶匙
五香粉 1 茶匙

3 tsps salt
1 tsp five spices powder

⟨⟨⟨ 做法 | Method

1. 雞翼洗淨，瀝乾水分，加入醃料醃 2 小時，汆水，瀝乾水分。
2. 燒熱油鑊，放入雞翼煎至金黃備用。
3. 暖油鑊，將蒜茸炸至金黃色，加入調味料拌勻，盛起，隔去油分。
4. 將炸蒜茸鋪在雞翼上即可上碟。

1. Rinse chicken wings and drain. Marinate for 2 hours, scald and drain.
2. Heat wok with oil, fry chicken wings until golden, set aside.
3. Warm wok with oil, deep-fry minced garlic until golden, add in seasonings and stir well. Dish up and drain excess oil.
4. Strew deep-fried garlic onto chicken wings and serve.

露斯瑪莉牛油雞翼

Fried Chicken Wings with Rosemary and Butter

⟳ 材料 | Ingredients

雞中翼 8 隻
橄欖油 1 湯匙
牛油（軟化）1 湯匙

8 chicken mid-joint wings
1 tbsp olive oil
1 tbsp melted butter

2~4 人
Serves 2~4

10~15 分鐘
10~15 minutes

◯◯◯ 醃料 | Marinade

露斯瑪莉 2 茶匙
生抽 2 茶匙
糖 1 茶匙
鹽 1/2 茶匙

2 tsps rosemary
2 tsps light soy sauce
1 tsp sugar
1/2 tsp salt

◯◯◯ 做法 | Method

1. 雞翼洗淨，瀝乾水分，加入醃料醃 30 分鐘。
2. 燒熱鑊，下橄欖油和牛油，下雞翼煎至金黃即成。

1. Rinse chicken wings well and drain. Marinate for 30 minutes.
2. Heat wok with oil, add olive oil and butter, put in chicken wings and fry until golden. Serve.

比翼雙飛 — Stuffed Chicken Wings

⦿⦿ 材料 | Ingredients

雞中翼 600 克	600g chicken mid-joint wings
紅蘿蔔 1/2 條	1/2 stick carrot
西芹 2 棵	2 sticks celery
冬菇 5 朵	5 dried black mushrooms
火腿（厚）1 片	1 slice ham (thick)

⦿⦿ 雞翼醃料 | Marinade for chicken wings

生抽 2 1/2 茶匙	1/2 tsps light soy sauce
鹽 1 茶匙	1 tsp salt
糖 1 茶匙	1 tsp sugar
酒 1 茶匙	1 tsp wine
胡椒粉 1/2 茶匙	1/2 tsp pepper powder

⊗ 冬菇醃料 | Marinade for mushrooms

生抽 1 茶匙	1 tsp light soy sauce
糖 1/2 茶匙	1/2 tsp sugar

⊗ 芡汁 | Thickening

蠔油 1 湯匙	1 tbsp oyster sauce
糖 2 茶匙	2 tsps sugar
老抽 1 1/2 茶匙	1 1/2 tsps dark soy sauce
生抽 1 茶匙	1 tsp light soy sauce

⊗ 做法 | Method

1. 雞翼洗淨，瀝乾水分，去骨，加入雞翼醃料醃 30 分鐘。
2. 火腿、紅蘿蔔、西芹洗淨，切粗條，約 5 厘米長。
3. 冬菇洗淨，浸軟，去蒂，切粗條，加冬菇醃料略醃。
4. 燒熱油鑊，加入火腿、紅蘿蔔、西芹和冬菇略炒，下鹽拌勻，盛起。
5. 將配料釀入雞翼內，用油半煎炸至金黃色，盛起。
6. 燒熱油鑊，加入芡汁煮滾，倒入雞翼拌勻即成。

1. Rinse chicken wings well, drain and remove bones. Marinate for 30 minutes.
2. Rinse ham, carrot and celery well, cut into thick strips in about 5 cm long.
3. Rinse mushrooms well, soak until soft, remove stalks, cut into thick strips. Add in marinade.
4. Heat wok with oil, put in ham, carrot, celery and mushrooms and stir-fry briefly, add salt and mix well. Dish up.
5. Stuff trimmings into chicken wings, deep-fry with little oil until golden brown. Dish up.
6. Heat wok with oil, bring sauce to a boil, add in chicken wings and stir well. Serve.

入廚貼士 | Cooking Tips

- 西芹可以西蘭花的梗或粟米芯代替。
- Celery could be replaced by broccoli's stems or baby corn shoots.

百里香吞拿魚釀雞翼

Fried Stuffed Chicken Wings with Tuna and Thyme

<CCO> **材料 | Ingredients**

雞全翼 8 隻
白酒 1/2 杯
雞湯 1/2 杯

8 chicken whole wings
1/2 cup white wine
1/2 cup stock

2~4 人
Serves 2~4

30~35 分鐘
30~35 minutes

餡料 | Fillings

吞拿魚 100 克
蒜茸 2 湯匙
百里香 2 茶匙

100g tuna
2 tbsps minced garlic
2 tsps thyme

醃料 | Marinade

生抽 1/2 湯匙
糖 1 茶匙
生粉 1/2 茶匙
油 1/2 茶匙

1/2 tbsp light soy sauce
1 tsp sugar
1/2 tsp caltrop starch
1/2 tsp oil

做法 | Method

1. 雞翼洗淨，瀝乾水分，切去雞槌，雞中翼和雞翼尖留用，去骨，加入醃料醃 30 分鐘。

2. 將餡料拌勻，釀入雞翼中，用竹籤縫上接口。

3. 燒熱油鑊，下雞翼煎至表面金黃，灒白酒，加入雞湯，轉慢火煮至汁液濃稠、雞翼全熟，拆去竹籤即可上碟。

1. Rinse chicken wings well and drain. Cut away drumsticks and keep mid-joints and wing tips for later use, remove bones. Marinate for 30 minutes.

2. Stir well the fillings, stuff into chicken wings. Seal with bamboo sticks.

3. Heat wok with oil, put in chicken wings and fry until the surface turns into golden. Sizzle with white wine, add in stock, switch to low heat and cook until the sauce thickens and chicken wings are done. Remove bamboo sticks and serve.

印尼炸雞翼

Indonesian Deep-fried Chicken Wings

◯◯◯ 材料 | Ingredients

雞中翼 8 隻
酸子 8 粒
水 1 1/2 杯

8 chicken mid-joint wings
8 cloves tamarind
1 1/2 cups water

2~4 人
Serves 2~4

10~15 分鐘
10~15 minutes

醃料 | Marinade

乾葱茸 1 1/2 湯匙
蒜茸 1 湯匙
鹽 1 茶匙
芫荽粉 1 茶匙
小茴粉 1/2 茶匙
黃薑粉 1/2 茶匙
胡椒粉 1/4 茶匙

1 1/2 tbsps minced shallot
1 tbsp minced garlic
1 tsp salt
1 tsp coriander powder
1/2 tsp cumin seed powder
1/2 tsp yellow ginger powder
1/4 tsp pepper powder

做法 | Method

1. 雞翼洗淨，瀝乾水分，加入醃料醃 30 分鐘。
2. 酸子用水浸片刻，去核，將肉連汁拌入雞翼內醃約 1 小時。
3. 燒熱油鑊，放入雞翼炸至金黃香脆即成。

1. Rinse chicken wings, drain and marinate for 30 minutes.
2. Soak tamarinds for a while, remove seeds, add flesh and juice of tamarinds into chicken wings and marinate for about 1 hour.
3. Heat wok with oil, put in chicken wings and deep-fry until golden brown and crispy. Serve.

香葱花雕雞翼

Drunken Chicken Wings with Spring Onion

⟨○○⟩ 材料 | Ingredients

雞中翼 600 克
葱段 80 克
薑 4 片乾
葱茸 2 湯匙

600g chicken mid-joint wings
80g spring onion sections
4 slices ginger
2 tbsps minced shallot

2~4 人
Serves 2~4

10~15 分鐘
10~15 minutes

醃料 | Marinade

老抽 1 湯匙
鹽 1 茶匙
糖 1 茶匙

1 tbsp dark soy sauce
1 tsp salt
1 tsp sugar

入廚貼士 | Cooking Tips
- 使用瓦煲燜焗效果較佳。
- Using an earthenware pot when stewing would be better.

汁料 | Sauce

花雕酒 2 湯匙
沙薑粉 2 湯匙
八角 4 粒
葱茸 2 湯匙
薑茸 2 湯匙
水適量

2 tbsps Shaoxing wine
2 tbsps ginger spice powder
4 cloves star anise
2 tbsps minced spring onion
2 tbsps minced ginger
water

做法 | Method

1. 雞翼洗淨，瀝乾水分，加入醃料醃 30 分鐘。
2. 燒熱油鑊，爆香薑、葱和乾葱頭，放入雞翼煎至金黃，加入汁料，加蓋燜約 10 分鐘至熟，即可上碟。

1. Rinse chicken wings well and drain. Marinate for 30 minutes.
2. Heat wok with oil, stir-fry ginger, spring onion and shallot until fragrant. Put in chicken wings and fry until golden. Add in sauce, cover and stew for 10 minutes until done. Serve.

鵝肝醬煎釀香雞翼

Deep-fried Stuffed Chicken Wings with Foie Gras

◯◯◯ 材料 | Ingredients

雞全翼 8 隻
椒鹽少許（隨意）

8 chicken whole wings
spicy pepper salt (at will)

2~4 人
Serves 2~4

30~35 分鐘
30~35 minutes

⦵ 餡料 | Fillings

法國鵝肝醬 100 克
西芹 40 克

100g foie gras
40g celery

⦵ 醃料 | Marinade

白酒 1 茶匙
五香粉 1/2 茶匙
胡椒粉少許

1 tsp white wine
1/2 tsp five spices powder
pepper powder

⦵ 做法 | Method

1. 雞翼洗淨，瀝乾水分，切去雞槌，雞中翼和雞翼尖留用，去骨。
2. 加入醃料醃 30 分鐘。
3. 西芹洗淨，切幼粒，與鵝肝醬拌勻。
4. 將餡料釀入雞翼內，用竹籤縫上接口。
5. 燒熱大鍋油，放入雞翼炸約 2 分鐘，改慢火再炸 2 分鐘，再轉大火炸至金黃色，盛起，拆去竹籤，上碟即可。

1. Rinse chicken wings well and drain. Cut away drumsticks, keep mid-joints and wing tips for later use. Remove bones.
2. Marinate for 30 minutes.
3. Wash celery well, cut into small dice, stir well with foie gras.
4. Stuff fillings into chicken wings. Seal with bamboo sticks.
5. Heat wok with large amount of oil, put in chicken wings and deep-fry for 2 minutes. Switch to low heat and deep-fry for 2 minutes again. Then switch to high heat and deep-fry until golden. Dish up and remove the bamboo sticks, serve.

醉香雞翼

Steamed Chicken Wings
with Wine

⬤⬤ 材料 | Ingredients

雞中翼 8 隻
鹽 1/2 湯匙

8 chicken mid-joint wings
1/2 tbsp salt

香噴噴 ~ Fragrant

◯◯◯ 汁料 | Sauce

雞汁（蒸雞的汁）1 杯
水 1 1/2 杯
紹酒 1 1/2 杯
糖 1 茶匙

1 cup chicken sauce
(sauce from steaming chicken)
1 1/2 cups water
1 1/2 cups Shaoxing wine
1 tsp sugar

入廚貼士 | Cooking Tips

* 此菜式適宜冷吃，汁液可留待下次再用。
* It is better to serve this dish in cold, chicken sauce can be kept for the later use.

◯◯◯ 做法 | Method

1. 雞翼洗淨，瀝乾水分，用鹽擦勻，醃約 4 小時。
2. 雞翼放碟上，隔水蒸約 15 分鐘至熟，隔出汁液備用。
3. 將汁料煮滾，盛深盆中，放下雞翼浸至汁料變冷，再放雪櫃中 6 小時（浸着汁料）。

1. Rinse chicken wings and drain. Dredge with salt evenly and marinate for 4 hours.
2. Arrange chicken wings on a plate, steam for 15 minutes until done. Drain and reserve the sauce, set aside.
3. Bring sauce to a boil, pour into a deep bowl, add in chicken wings and soak until cold, put into refrigerator for 6 hours.

麥片香草脆雞翼

Deep-fried Chicken Wings with Oatmeal and Herbs

2~4 人
Serves 2~4

20~25 分鐘
20~25 minutes

材料 | Ingredients

雞中翼 8 隻
雞蛋 1 隻
麥片 5 湯匙
麵粉 4 湯匙
牛油（軟化）2 湯匙
蒜茸 1 茶匙
香草 1 茶匙

8 chicken mid-joint wings
1 egg
5 tbsps oatmeal
4 tbsps flour
2 tbsps melted butter
1 tsp minced garlic
1 tsp herbs

醃料 | Marinade

生抽 1 茶匙
酒 1 茶匙
鹽 1/2 茶匙
胡椒粉少許

1 tsp light soy sauce
1 tsp wine
1/2 tsp salt
pepper powder

做法 | Method

1. 雞翼洗淨，瀝乾水分，加入醃料醃 10 分鐘。
2. 蒜茸與牛油拌勻。雞蛋打勻。
3. 雞翼以蒜茸牛油塗勻，撲上麵粉，沾上蛋液，再均勻地滾上麥片。
4. 燒熱油鑊，下雞翼中炸至金黃，撒上香草即成。

1. Rinse chicken wings well and drain. Marinate for 10 minutes.
2. Mix garlic and butter well. Beat egg.
3. Dredge chicken wings with garlic butter, coat with flour and beaten egg. Then coat with oatmeal evenly.
4. Heat wok with oil, deep-fry chicken wings until golden. Strew with oregano and serve.

野菌釀雞翼配香草汁

Stuffed Chicken Wings with Mushrooms in Sauce with Herbs

材料 | Ingredients

雞全翼 8 隻

8 chicken whole wings

餡料 | Fillings

日本冬菇 6 朵
牛肝菌 6 朵
雞髀菇 1 個
洋蔥 1/2 個

6 Japanese mushrooms
6 boletus calopus
1 chicken leg mushrooms
1/2 onion

2~4 人
Serves 2~4

30~35 分鐘
30~35 minutes

⟨⟨⟩⟩ 調味料 | Seasonings

黑松露菌醬 1 湯匙
鹽 1/2 茶匙
胡椒粉少許

1 tbsp black truffle pate
1/2 tsp salt
pepper powder

⟨⟨⟩⟩ 汁料 | Sauce

蠔油 2 湯匙
生抽 1 茶匙
老抽 1 茶匙
糖 1 茶匙
香草 1/2 茶匙

2 tbsps oyster sauce
1 tsp light soy sauce
1 tsp dark soy sauce
1 tsp sugar
1/2 tsp herbs

⟨⟨⟩⟩ 做法 | Method

1. 雞翼洗淨，瀝乾水分，切去雞槌，雞中翼和雞翼尖留用，去骨。
2. 冬菇洗淨，去蒂，切碎；洋葱去衣，切碎；牛肝菌、雞髀菇洗淨，切碎。
3. 燒熱油鑊，將餡料炒香，下調味料拌勻，盛起。
4. 將餡料釀入雞翼中，用竹籤縫上接口。
5. 燒熱油鑊，下雞翼煎至金黃，加入汁料，煮至濃稠，拆去竹籤即成。

1. Rinse chicken whole wings and drain. Cut away drumsticks, keep mid-joints and wing tips for later use, remove bones.
2. Wash mushrooms and remove stalks, chop finely. Peel onion and chop finely. Wash boletus calopus and chicken leg mushrooms well, chop finely.
3. Heat wok with oil, stir-fry fillings until fragrant. Add in seasoning and stir well, dish up.
4. Stuff fillings into chicken wings. Seal with bamboo sticks.
5. Heat wok with oil, put in chicken wings and fry till golden. Add in sauce and cook until thick. Remove bamboo sticks and serve.

Crispy Chicken Wings with Tea Leaves

脆皮茶香雞翼

2~4 人
Serves 2~4

20~25 分鐘
20~25 minutes

⟪⟫ 材料 | Ingredients

雞中翼 600 克
薑 3 片
蔥（切段）2 棵
八角 2 粒
花椒 1/2 茶匙
桂皮 1 塊
普洱茶葉 1 1/2 湯匙

600g chicken mid-joint wings
3 slices ginger
2 stalks spring onion (sectioned)
2 cloves star anise
1/2 tsp Chinese red pepper
1 piece Chinese cinnamon
1 1/2 tbsps Pu Er tea leaves

⟪⟫ 調味料 | Sauce

生抽 1/3 杯
老抽 1/4 杯
紹酒 3 湯匙
冰糖碎 2 湯匙
水 1 1/2 杯

1/3 cup light soy sauce
1/4 cup dark soy sauce
3 tbsps Shaoxing wine
2 tbsps rock sugar (crushed)
1 1/2 cup water

⟪⟫ 做法 | Method

1. 雞翼洗淨，瀝乾水分。
2. 燒熱油鑊，爆香薑、蔥，加入八角、花椒、桂皮和茶葉，下調味料煮約 15 分鐘。
3. 放入雞翼，煮約 5 分鐘，熄火，浸約 4 小時，盛起待涼。
4. 再將雞翼放油鑊中炸至金黃即可。

1. Rinse chicken wings and drain.
2. Heat wok with oil, stir-fry ginger, spring onion until fragrant. Add in star anise, Chinese red pepper, Chinese cinnamon and tea leaves. Add seasonings and cook for 15 minutes.
3. Put in chicken wings and cook for 5 minutes. Switch off the heat and soak for about 4 hours. Dish up and leave to cool.
4. Deep-fry chicken wings until golden. Serve.

啤酒燴雞翼

Braised Chicken Wings with Beer

◯◯◯ 材料 | Ingredients

雞中翼 600 克	600g mid–joint chicken wings
洋葱 2 個	2 onions
青椒 1 個	1 green pepper
黃椒 1 個	1 yellow pepper
紅椒 1 個	1 red pepper
香葉 1 片	1 bay leaf
牛油 3 湯匙	3 tbsps butter
紅椒粉 1 茶匙	1 tsp cayenne
蒜頭 2 粒	2 cloves garlic

2~4 人
Serves 2~4

25~30 分鐘
25~30 minutes

醃料 | Marinade

胡椒粉 1 茶匙
鹽 1 茶匙

1 tsp pepper powde
1 tsp salt

汁料 | Sauce

啤酒 150 毫升
茄汁 3 湯匙
鹽 2 茶匙

150 ml beer
3 tbsps ketchup
2 tsps salt

做法 | Method

1. 雞翼洗淨，瀝乾水分，加入醃料醃 10 分鐘。

2. 洋蔥、青椒、黃椒、紅椒洗淨，去籽，分別切塊。

3. 燒熱鑊，以慢火煮融牛油，將雞翼煎至金黃，盛起。

4. 再燒熱鑊，下油爆香蒜頭，棄去，倒入雞翼、洋蔥、香葉、紅椒粉拌勻，下汁料拌勻，加蓋燜 20 分鐘。倒入其餘材料煮一會兒，棄去香葉，即可上碟。

1. Rinse chicken wings and drain. Marinate for 10 minutes.

2. Wash onion, green pepper, yellow pepper and red pepper, remove the seeds and chop into pieces.

3. Heat wok and melt butter over low heat. Fry chicken wings until golden and dish up.

4. Heat wok with oil again, stir-fry garlic until fragrant, discard garlic. Put in chicken wings, onion, bay leaf and cayenne and mix well. Add in sauce and stir well. Cover and stew for 20 minutes. Put in the remaining ingredients and cook for a while, discard bay leaf and serve.

香滑芝士焗雞翼

Baked Chicken Wings with Cheese

◯◯◯ 材料 | Ingredients

雞中翼 8 隻
芝士茸 3 湯匙

8 chicken mid–joint wings
3 tbsps minced cheese

醃料 | Marinade

生抽 1 1/2 茶匙
糖 1 茶匙
酒 1 茶匙
生粉 1 茶匙
鹽 1/2 茶匙

1 1/2 tsps light soy sauce
1 tsp sugar
1 tsp wine
1 tsp caltrop starch
1/2 tsp salt

做法 | Method

1. 雞翼洗淨，瀝乾水分，加入醃料醃 1 小時。

2. 燒熱油鑊，放入雞翼煎至金黃。

3. 預熱焗爐至 200℃，將雞翼排放在已塗油的焗盤上，灑上芝士茸，焗約 10 分鐘至金黃即可。

1. Rinse chicken wings well and drain. Marinate for 1 hour.

2. Heat wok with oil, put in chicken wings and fry until golden.

3. Preheat oven to 200℃ . Arrange chicken wings onto a greased baking tray. Strew minced cheese on the top and bake for 10 minutes until golden, serve.

2~4 人
Serves 2~4

15~20 分鐘
15~20 minutes

蜜糖香茅雞翼

Deep-fried Chicken Wings with Honey and Lemongrass

◯◯◯ 材料 | Ingredients

雞中翼 8 隻

8 chicken mid-joint wings

⊗⊗ 醃料 | Marinade

生抽 1 湯匙
蜜糖 1 湯匙
蒜粉 2 茶匙
香茅粉 2 茶匙
酒 1 茶匙
鹽 1/2 茶匙
胡椒粉少許

1 tbsp light soy sauce
1 tbsp honey
2 tsps garlic powder
2 tsps lemongrass powder
1 tsp wine
1/2 tsp salt
pepper power

⊗⊗ 做法 | Method

1. 雞翼洗淨，瀝乾水分，加入醃料醃 1 小時。
2. 燒熱油鑊，下雞翼炸至金黃，盛起隔油即成。

1. Rinse chicken wings well and drain. Marinate for
   1 hour.
2. Heat wok with oil, deep-fry chicken wings until
   golden, dish up and drain excess oil. Serve.

2~4 人
Serves 2~4

20~25 分鐘
20~25 minutes

鮮果皮茸洋蔥雞翼

Stewed Chicken Wings with Fresh
Fruit Rind and Onion

⨀ 材料 | Ingredients

雞中翼 600 克
洋蔥 2 個
老抽 1 湯匙
薑茸 1/2 湯匙
檸檬皮茸 1/2 湯匙
鮮橙皮茸 1/2 湯匙

600g chicken mid-joint wings
2 onions
1 tbsp dark soy sauce
1/2 tbsp minced ginger
1/2 tbsp ground lemon rind
1/2 tbsp ground orange rind

◯◯◯ 醃料 | Marinade

紹酒 1 湯匙
老抽 1 湯匙
糖 1 茶匙
鹽 1/2 茶匙

1 tbsp Shaoxing wine
1 tbsp dark soy sauce
1 tsp sugar
1/2 tsp salt

◯◯◯ 汁料 | Sauce

生粉 1/2 茶匙
麻油少許
胡椒粉少許
水 150 毫升

1/2 tsp caltrop starch
sesame oil
pepper powder
150ml water

◯◯◯ 做法 | Method

1. 雞翼洗淨，瀝乾水分，加入醃料醃 30 分鐘。以老抽搽勻雞翼外皮。
2. 洋葱去衣，切件。
3. 燒熱油鑊，下雞翼炸至金黃，盛起隔油。
4. 燒熱油 2 湯匙，爆香洋葱、薑茸及鮮果皮茸，將雞翼放回鑊，加入汁料，煮滾後以慢火燜至汁料濃稠即可上碟。

1. Rinse chicken wings well and drain. Marinate for 30 minutes. Strew chicken wings with dark say sauce evenly.
2. Peel onion and cut into pieces.
3. Heat wok with oil and deep-fry chicken wings until golden. Drain excess oil and dish up.
4. Heat wok with 2 tbsps oil, stir-fry onion, minced ginger, ground lemon rind and orange rind. Return chicken wings to wok, add in sauce and bring it to a boil. Stew over low heat until sauce thickens and serve.

可樂雞翼

Stewed Chicken Wings with Cola

2~4 人
Serves 2~4

15~20 分鐘
15~20 minutes

⊙⊙ 材料 | Ingredients

雞中翼 600 克	600g chicken mid-joint wings
蒜頭 2 粒	2 cloves garlic
薑 1 片	1 slice ginger
可樂 1 罐	1 can cola

⊙⊙ 醃料 | Marinade

乾葱茸 2 湯匙	2 tbsps minced shallot
老抽 2 湯匙	2 tbsps dark soy sauce
生抽 2 茶匙	2 tsps light soy sauce
鹽 1 茶匙	1 tsp salt
喼汁 1 茶匙	1 tsp Worcestershire sauce
冧酒 1 茶匙	1 tsp rum
糖 1/2 茶匙	1/2 tsp sugar
胡椒粉 1/2 茶匙	1/2 tsp pepper powder

⊙⊙ 做法 | Method

1. 雞翼洗淨，瀝乾水分，加入醃料醃 1 小時。
2. 蒜頭去衣，與薑片分別略拍，備用。
3. 燒熱油鑊，爆香蒜頭、薑片，加入雞翼煎至金黃，倒入可樂，加蓋燜約 15 分鐘至汁液濃稠即可。

1. Rinse chicken wings well and drain. Marinate for 1 hour.
2. Peel and crush garlic and crush ginger, set aside.
3. Heat wok with oil, stir-fry garlic and ginger until fragrant. Fry chicken wings until golden. Pour in cola, cover and cook for 15 minutes until sauce thickens. Serve.

果醬香煎雞翼

Fried Chicken Wings
with Apricot Jam

◯◯ 材料 | Ingredients

雞中翼 8 隻
薑 3 片
紹酒 1 茶匙

8 chicken mid-joint wings
3 slices ginger
1 tsp Shaoxing wine

2~4 人
Serves 2~4

10~15 分鐘
10~15 minutes

◯◯ 醃料 | Marinade

生抽 1 茶匙
老抽 1 茶匙
酒 1 茶匙
胡椒粉少許

1 tsp light soy sauce
1 tsp dark soy sauce
1 tsp wine
pepper powder

◯◯ 調味料 | Seasonings

黃梅果醬 4 湯匙
生抽 2 茶匙
鹽 1/4 茶匙
水 1/2 杯

4 tbsps apricot jam
2 tsps light soy sauce
1/4 tsp salt
1/2 cup water

◯◯ 做法 | Method

1. 雞翼洗淨，瀝乾水分，加入醃料醃 20 分鐘。
2. 燒熱油鑊，下雞翼煎至金黃，盛起隔油。
3. 燒熱油，爆香薑片，下雞翼爆透，潷酒，加入調味料，煮滾後用中慢火煮至汁液濃稠即成。

1. Rinse chicken wings well and drain. Marinate for 20 minutes.
2. Heat wok with oil, fry chicken wings until golden, drain excess oil and dish up.
3. Heat wok with oil, stir-fry ginger until fragrant. Put in chicken wings and stir-fry until done. Add in seasonings and bring to a boil. Cook over medium-low heat until sauce thickens, serve.

泰式九層塔雞翼尖

Fried Chicken Wing Tips with Basil in Thai Style

材料 | Ingredients

雞翼尖 300 克
300g chicken wing tips

醃料 | Marinade

生抽 1 茶匙
生粉 1 茶匙
香茅（切碎）1/2 枝

1 tsp light soy sauce
1 tsp caltrop starch
1/2 stick chopped lemongrass

2~4 人
Serves 2~4

15~20 分鐘
15~20 minutes

◯◯◯ **汁料 | Sauce**

泰式甜辣醬 4 湯匙
九層塔 10 塊

4 tbsps Thai sweet and spicy sauce
10 pieces basil

◯◯◯ **做法 | Method**

1. 雞翼尖洗淨，瀝乾水分，加入醃料醃
   30 分鐘。
2. 燒熱油鑊，將雞翼尖煎熟，加入泰式
   甜醬略煮，再加九層塔拌勻即可。

1. Rinse chicken wing tips well and
   drain. Marinate for 30 minutes.
2. Heat wok with oil, fry chicken
   wing tips until done. Add in Thai
   sweet and spicy sauce and cook
   for a while. Put in basil and stir
   well, serve.

入廚貼士 | Cooking Tips
- 九層塔又名金不換，不可炒得太久。
- Basil could not be cooked for too long.

味噌南瓜魚乾雞翼尖

Stewed Wing Tips with Miso, Pumpkin and Dried Fish

◯◯◯ 材料 | Ingredients

雞翼尖 300 克
南瓜 300 克
魚乾 2 湯匙

300g chicken wing tips
300g pumpkin
2 tbsps dried fish

2~4 人
Serves 2~4

20~25 分鐘
20~25 minutes

◯◯ 醃料 | Marinade

生抽 1 茶匙
糖 1 茶匙
薑茸 1 茶匙
老抽 1/2 茶匙

1 tsp light soy sauce
1 tsp sugar
1 tsp minced ginger
1/2 tsp dark soy sauce

◯◯ 汁料 | Sauce

味噌 1 茶匙
糖 1/2 茶匙
水 150 毫升

1 tsp miso
1/2 tsp sugar
150ml water

◯◯ 做法 | Method

1. 雞翼尖洗淨,瀝乾水分,加入醃料醃 10 分鐘。
2. 燒熱油鑊,下雞翼尖煎至金黃,盛起隔油。
3. 原鑊下南瓜和魚乾略炒,下雞翼尖和汁料拌勻,加蓋,以中火燜至南瓜變軟即可。

1. Rinse chicken wings well and drain. Marinate for 10 minutes.
2. Heat wok with oil, fry chicken wing tips until golden. Dish up and drain excess oil.
3. Stir–fry pumpkin and dried fish in the original wok. Put in chicken wing tips and sauce and stir–fry for a while. Cover and stew over medium heat until pumpkin becomes soft. Serve.

日式清酒焗香雞翼

Japanese Baked Chicken Wings with Sake

2~4 人
Serves 2~4

20~25 分鐘
20~25 minutes

材料 | Ingredients

雞中翼 8 隻
蜜糖適量
8 chicken mid-joint wings
honey

醃料 | Marinade

日本清酒 2 杯
生抽 1 杯
糖 3 湯匙

2 cups sake
1 cup light soy sauce
3 tbsps sugar

做法 | Method

1. 雞翼洗淨，瀝乾水分，加入醃料拌勻，放入雪櫃冷藏一晚。

2. 預熱焗爐至 200℃。

3. 將錫紙掃油，放上雞翼，包好，放入焗爐焗 15 分鐘，取出，塗上蜜糖再焗 5 分鐘即成。

1. Rinse chicken wings well and drain, add in marinade and stir well, put into the refrigerator for one night.

2. Preheat an oven to 200℃ .

3. Strew oil onto a piece and aluminium foil, arrange chicken wings on it and wrap properly. Bake in oven for 15 minutes. Take out and strew with honey and bake for 5 more minutes, serve.

蜜汁梅子洋蔥雞翼

Stir-fried Chicken Wings with Plum and Onion in Honey Sauce

材料 | Ingredients

雞中翼 8 隻
梅子（去核）3 粒
薑絲 1 湯匙
洋蔥 1/2 個

8 chicken mid-joint wings
3 plums (cored)
1 tbsp shredded ginger
1/2 onion

2~4 人
Serves 2~4

20~25 分鐘
20~25 minutes

醃料 | Marinade

酒 1 茶匙
生抽 1 茶匙
糖 1/2 茶匙
鹽 1/2 茶匙

1 tsp wine
1 tsp light soy sauce
1/2 tsp sugar
1/2 tsp salt

調味料 | Seasonings

蜜糖 3 湯匙
老抽 1 湯匙
水 5 湯匙

3 tbsps honey
1 tbsp dark soy sauce
5 tbsps water

做法 | Method

1. 雞翼洗淨，瀝乾水分，加入醃料醃 1 小時。

2. 洋葱去衣，切絲。

3. 燒熱鑊，下油 1 湯匙，放入雞翼煎至兩面金黃色，盛起。

4. 再燒熱鑊，下油爆香薑絲和洋葱，將雞翼回鑊，加入梅子肉和調味料，煮至雞翼全熟及汁液濃稠即成，上碟。

1. Rinse chicken wings well and drain. Marinate for 1 hour.

2. Peel onion and shred.

3. Heat wok with 1 tbsp of oil, fry chicken wings until both sides turn golden, dish up.

4. Heat wok with oil again, stir-fry ginger and onion until fragrant. Return chicken wings. Add in plums and seasonings and cook until the sauce thickens, serve.

南乳燒雞翼

Deep-fried Chicken Wings with
Fermented Tarocurd

⭕⭕ 材料 | Ingredients

雞中翼 8 隻
蜜糖 2 茶匙
水 5 湯匙

8 chicken mid-joint wings
2 tsps honey
5 tbsps water

⊘⊘ **調味料 | Seasonings**

乾葱頭 4 粒	4 cloves shallot
南乳 1/2 件	1/2 fermented tarocurd
玫瑰露 2 湯匙	2 tbsps rose wine
糖 1/2 湯匙	1/2 tbsp sugar
生抽 1 茶匙	1 tsp light soy sauce

⊘⊘ **做法 | Method**

1. 雞翼洗淨，瀝乾水分，汆水，過冷河。

2. 拌勻調味料，均勻地塗在雞翼上。

3. 蜜糖用清水煮融，塗在雞翼上，待乾備用。

4. 燒熱大鍋油，放入雞翼，以慢火炸至金黃全熟即成。

1. Rinse chicken wings well and drain. Scald and rinse.

2. Stir well seasonings and dredge onto chicken wings evenly.

3. Melt honey in water, dredge onto chicken wings, set aside and leave to dry.

4. Heat wok with large amount of oil, deep-fry chicken wings over low heat until golden and done, serve.

蜜桃茸燒雞翼

Deep-fried Chicken Wings with Minced Peach

⟨⟨⟩⟩ 材料 | Ingredients

雞中翼 8 隻　　8 chicken mid-joint wings
蒜茸 2 茶匙　　2 tsps minced garlic

⟨⟨⟩⟩ 茨汁 | Thickening

生粉 2 茶匙　　2 tsps caltrop starch
水 1 湯匙　　　1 tbsp water

⊘⊘ 醃料 | Marinade

豉油雞汁 1 湯匙	1 tbsp chicken dark soy sauce
生粉 1 湯匙	1 tbsp caltrop starch
生抽 1 湯匙	1 tbsp light soy sauce
蒜粉 1/2 湯匙	1/2 tbsp garlic powder
鹽 1 茶匙	1 tsp salt
黑椒粉 1/2 茶匙	1/2 tsp black pepper powder

⊘⊘ 汁料 | Sauce

蜜桃茸 1/3 杯	1/3 cup minced peach
白酒 25 毫升	25 ml white wine
蜜桃糖水 25 毫升	25 ml peach syrup
茄汁 1 1/2 湯匙	1 1/2 tbsps ketchup
蠔油 3/4 湯匙	3/4 tbsp oyster sauce
喼汁 1 茶匙	1 tsp Worcestershire sauce
水 50 毫升	50 ml water

⊘⊘ 做法 | Method

1. 雞翼洗淨，瀝乾水分，加入醃料醃 1 小時。

2. 燒熱油鑊，下雞翼炸至金黃，盛起隔油，上碟。

3. 下油爆香蒜茸，倒入蜜桃茸和白酒煮片刻，下其他汁料，慢火煮滾，調生粉水勾芡，淋上雞翼上即成。

1. Rinse chicken wings well and drain. Marinate for 1 hour.

2. Heat wok with oil, deep-fry chicken wings until golden. Drain excess oil and arrange onto a plate.

3. Stir-fry minced garlic until fragrant, put in minced peach and white wine and cook for a while. Add the remaining sauce and bring to a boil over low heat. Put in thickening and mix well. Pour onto chicken wings and serve.

芝士釀雞翼

Stuffed Chicken Wings with Mozzarella

材料 | Ingredients

雞中翼 8 隻
意大利白芝士 100 克
麵包糠 1 杯
生粉 1 杯
雞蛋 1 隻
檸檬 1/2 個

8 chicken mid–joint wings
100g Mozzarella cheese
1 cup breadcrumbs
1 cup caltrop starch
1 egg
1/2 lemon

◎◎ 醃料 | Marinade

黑胡椒粉 1 茶匙
迷迭香碎 1/2 茶匙
芥辣 1/2 茶匙
鹽 1/2 茶匙

1 tsp black pepper powder
1/2 tsp chopped rosemary
1/2 tsp mustard
1/2 tsp salt

◎◎ 做法 | Method

1. 雞翼洗淨，瀝乾水分，去骨，加入醃料醃 30 分鐘。

2. 芝士切條，釀入雞翼內。

3. 雞蛋打勻，檸檬榨汁。

4. 預熱焗爐至 200℃。

5. 雞翼蘸上蛋液，撲上生粉，再蘸上蛋液，撲上麵包糠。放進焗爐內焗15分鐘至金黃色；翻轉另一面再焗10分鐘，灑上檸檬汁即成。

1. Rinse chicken wings well, drain and remove bones. Marinate for 30 minutes.

2. Shred Mozzarella cheese and stuff into chicken wings.

3. Beat egg well and juice lemon.

4. Preheat an oven to 200℃ .

5. Dip chicken wings in beaten egg and coat with caltrop starch. Then dip chicken wings in beaten egg and coat with breadcrumbs. Put into oven and bake for 15 minutes until golden. Turn over and bake for 10 minutes again, sprinkle with lemon juice.

2~4 人
Serves 2~4

15~20 分鐘
15~20 minutes

香芒雞翼

Deep-fried Chicken Wings
with Mango Sauce

⊙⊙ 材料 | Ingredients

雞中翼 600 克
芒果肉 1 個
青椒粒 2 湯匙
紅椒粒 2 湯匙
芒果醬 2 湯匙
生粉 2 湯匙
蒜茸 1 湯匙

600g chicken mid-joint wings
1 mango flesh
2 tbsps diced green pepper
2 tbsps diced red pepper
2 tbsps mango paste
2 tbsps caltrop starch
1 tbsp minced garlic

ⓒⓒ 醃料 | Marinade

生抽 1 1/2 湯匙	1 1/2 tbsps light soy sauce
乾葱茸 1 湯匙	1 tbsp minced shallot
糖 2 茶匙	2 tsps sugar
喼汁 1 1/2 茶匙	1 1/2 tsps Worcestershire sauce
鹽 1/2 茶匙	1/2 tsp salt
胡椒粉少許	pepper powder

ⓒⓒ 芡汁 | Thickening

生粉 1 茶匙	1 tsp caltrop starch
水 1 湯匙	1 tbsp water

ⓒⓒ 做法 | Method

1. 雞翼洗淨，瀝乾水分，加入醃料醃 30 分鐘。
2. 芒果肉切粒。
3. 燒熱油鑊，雞翼撲勻生粉，放鑊中炸至金黃熟透，盛起隔油上碟。
4. 下油爆香蒜茸，倒入芒果醬，一半果肉，勾芡，加入餘下果肉，淋在雞翼上即成。

1. Rinse chicken wings well and drain. Marinate for 30 minutes.
2. Cut mango flesh into dice.
3. Heat wok with oil, strew chicken wings with caltrop starch evenly. Deep-fry in wok until golden and done. Drain excess oil and arrange onto a plate.
4. Add oil in wok, stir-fry minced garlic until fragrant. Put in mango paste and half of the mange dice. Pour in thickening and cook for while, add in the remaining flesh and mix well. Pour onto the chicken wings, serve.

2~4 人
Serves 2~4

20~25 分鐘
20~25 minutes

話梅雞翼

Stir-fried Chicken Wings
with Preserved Prunes

◯◯◯ 材料 | Ingredients

雞中翼 8 隻	8 chicken mid–joint wings
話梅 6 粒	6 preserved prunes
中國芹菜 2 條	2 sticks Chinese celery
紅椒 1 隻	1 red pepper
洋葱 1 個	1 onion
薑 2 片	2 slices ginger
冰糖適量	rock sugar
酸梅 2 粒	2 sour plums
水 3 杯	3 cups water

⦾ 醃料 | Marinade

酒 1 湯匙
鹽 1/2 茶匙
胡椒粉少許

1 tbsp wine
1/2 tsp salt
pepper powder

⦾ 調味料 | Seasonings

甜豉油 1/2 湯匙
生粉 1 茶匙
水 1 湯匙

1/2 tbsp sweet soy sauce
1 tsp caltrop starch
1 tbsp water

⦾ 做法 | Method

1. 雞翼洗淨，瀝乾水分，加入醃料醃 30 分鐘。

2. 燒熱油鑊，下雞翼煎至金黃，盛起隔油備用。

3. 芹菜洗淨，切段；紅椒、洋葱洗淨，切粒。

4. 燒滾水，加入話梅、酸梅、一半紅椒，加蓋煮至話梅出味，盛起備用。

5. 燒熱油鑊，爆香薑片和洋葱，加入話梅汁、冰糖，將雞翼回鑊煮片刻，下芹菜、紅椒和調味料拌勻，即可上碟。

1. Rinse chicken wings well and drain. Marinate for 30 minutes.

2. Heat wok with oil, fry chicken wings until golden, dish up and drain excess oil.

3. Wash Chinese celery well, cut into sections. Wash red pepper and onion well, cut into dice.

4. Bring water to a boil, add in preserved prunes, sour plums and half of the red pepper dice. Cover and cook until the fragrance of preserved prunes come out. Dish up and set aside.

5. Heat wok with oil, stir-fry ginger and onion until fragrant. Add in preserved prune sauce and rock sugar, put in chicken wings and cook for a while. Add in Chinese celery, red pepper and seasonings, stir well and serve.

2~4 人
Serves 2~4

25~30 分鐘
25~30 minutes

紅酒燴雞槌

Stewed Drumsticks with Red Wine

材料 | Ingredients

雞槌 8 隻
蒜茸 1 湯匙
洋蔥 1 個
紅椒 1 個

8 chicken drumsticks
1 tbsp minced garlic
1 onion
1 red pepper

◯◯◯ 汁料 | Sauce

紅酒 1/2 杯	1/2 cup red wine
番茄茸 1/2 杯	1/2 cup minced tomatoes
水 1/2 杯	1/2 cup water
茄汁 150 毫升	150ml ketchup
百里香 1 茶匙	1 tsp thyme
鹽 1 茶匙	1 tsp salt
胡椒粉少許	pepper powder

◯◯◯ 做法 | Method

1. 雞槌洗淨，瀝乾水分。
2. 燒熱油鑊，下雞槌炸至金黃，盛起隔油備用。
3. 洋葱洗淨，去衣，切碎；紅西椒洗淨，去籽，切絲。
4. 燒熱油鑊，爆香蒜茸和洋葱，加入紅酒煮滾，加入番茄茸、茄汁和水拌勻。
5. 把雞槌、百里香加入鑊中，改用慢火煮 20 分鐘，再加入紅西椒、胡椒粉和鹽調味即可。

1. Rinse drumsticks well and drain.
2. Heat wok with oil, deep-fry drumsticks till golden. Dish up and drain excess oil, set aside.
3. Wash onion well, peel and dice. Wash red pepper well, remove seeds and shred.
4. Heat wok with oil, stir-fry minced garlic and onion until fragrant. Add in red wine and bring to a boil. Put in chopped tomatoes, ketchup and water and stir well.
5. Put drumsticks and thyme into the wok. Switch to low heat and cook for 20 minutes, put in red pepper, pepper powder and salt, serve.

香檸牛油雞翼

Fried Chicken Wings with Lemon and Butter

材料 | Ingredients

雞中翼 8 隻
牛油 1 湯匙
檸檬 1/2 個

8 chicken wings
1 tbsp butter
1/2 lemon

2~4 人
Serves 2~4

15~20 分鐘
15~20 minutes

醃料｜Marinade

生抽 1 湯匙
酒 1 茶匙
胡椒粉少許

1 tbsp light soy sauce
1 tsp wine
pepper powder

調味料｜Seasonings

檸檬汁 1 1/2 湯匙
糖 1/2 湯匙
鹽 1/4 茶匙

1 1/2 tbsps lemon juice
1/2 tbsp sugar
1/4 tsp salt

做法｜Method

1. 雞翼洗淨，瀝乾水分，加入醃料醃 30 分鐘。
2. 燒熱油鑊，下雞翼炸至金黃，盛起隔油備用。
3. 檸檬皮磨茸，檸檬肉榨汁備用。
4. 燒熱油鑊，將牛油以小火煮融，放下檸檬皮茸，加入調味料，下雞翼拌勻至均勻地沾上汁液即成。

1. Rinse chicken wings well and drain. Marinate for 30 minutes.
2. Heat wok with oil, deep-fry chicken wings until golden, dish up and drain excess oil. Set aside.
3. Grind lemon rind and squeeze lemon flesh into juice.
4. Heat wok with oil, melt butter over low heat. Put in ground lemon rind, seasonings and chicken wings, stir-fry well until the chicken wings are coated with sauce evenly. Serve.

意大利檸檬黑椒雞翼

Italian Baked Chicken Wings with Lemon and Black Pepper

材料 | Ingredients

雞中翼 8 隻
檸檬 3 個
香葉 2 片
迷迭香 2 棵
橄欖油 1 湯匙

8 chicken mid-joint wings
3 lemons
2 slices bay leaf
2 sticks rosemary
1 tbsp olive oil

2~4 人
Serves 2~4

30~35 分鐘
30~35 minutes

醃料 | Marinade

檸檬汁 1 個量
黑胡椒粉 2 茶匙
鹽 1 茶匙

1 lemon (juiced)
2 tsps whole black pepper
1 tsp salt

調味料 | Seasonings

牛油 75 克
番荽碎 2 湯匙
乾蔥茸 2 湯匙
蒜茸 1 湯匙

75g butter
2 tbsps chopped parsley
2 tbsps minced shallot
1 tbsp minced garlic

做法 | Method

1. 檸檬皮磨茸後，將其中 1 個榨汁，其餘切片。
2. 雞翼洗淨，瀝乾水分，加入醃料醃 30 分鐘。
3. 將調味料拌勻，均勻地塗在雞翼表面。
4. 預熱焗爐至 200℃。
5. 將剩餘的檸檬片和迷迭香放在雞翼上，掃上橄欖油，置焗爐焗 20–30 分鐘至金黃色即成。

1. Grind lemon rind, juice one of the lemons and slice the remaining lemons.
2. Rinse chicken wings well and drain. Marinate for 30 minutes.
3. Stir well seasonings, strew onto chicken wings evenly.
4. Preheat an oven to 200℃ .
5. Sprinkle the remaining lemon slices and rosemary onto chicken wings and strew with olive oil. Bake for 20–30 minutes until golden.

四川怪味雞翼

Sichuan Special Flavoured Chicken Wings

⊘⊘ 材料 | Ingredients

雞中翼 8 隻

8 chicken mid–joint wings

醃料 | Marinade

酒 1 茶匙	1 tsp wine
鹽 1/2 茶匙	1/2 tsp salt
胡椒粉少許	pepper powder

調味料 | Seasonings

蒜茸 2 湯匙	2 tbsps minced garlic
葱茸 2 湯匙	2 tbsps minced spring onion
薑茸 1 湯匙	1 tbsp minced ginger
麻油 1 湯匙	1 tbsp sesame oil
老抽 1 湯匙	1 tbsp dark soy sauce
芝麻醬 1 湯匙	1 tbsp sesame paste
糖 1 湯匙	1 tbsp sugar
黑醋 1 湯匙	1 tbsp dark vinegar
辣油 1 湯匙	1 tbsp chili oil
花椒 1 茶匙	1 tsp Chinese red pepper

做法 | Method

1. 雞翼洗淨，瀝乾水分，加入醃料醃 30 分鐘。
2. 將雞翼隔水蒸熟備用。
3. 將調味料拌勻，煮滾，淋於雞翼上便可。

1. Rinse chicken wings well and drain. Marinate for 30 minutes.
2. Steam chicken wings until done, set aside.
3. Stir well all the seasonings and bring to a boil, pour on the chicken wings and serve.

百變雞翼

編著
梁燕

編輯
Pheona Tse

美術設計
Venus Lo

排版
Wing Yeung

翻譯
劉海雯

攝影
Fanny

出版者
萬里機構出版有限公司
香港鰂魚涌英皇道1065號東達中心1305室
電話：2564 7511
傳真：2565 5539
電郵：info@wanlibk.com
網址：http://www.wanlibk.com
　　　http://www.facebook.com/wanlibk

發行者
香港聯合書刊物流有限公司
香港新界大埔汀麗路36號
中華商務印刷大廈3字樓
電話：2150 2100
傳真：2407 3062
電郵：info@suplogistics.com.hk

承印者
美雅印刷製本有限公司

出版日期
二零一八年四月第一次印刷

萬里機構

萬里 Facebook